The A's & B's of Our Inland Sea

Written by
Nicole Anderson

Illustrated by
Johanna Bossart

The A's & B's of Our Inland Sea

May you find your love for wild places and magical adventures at our inland sea, our Great Salt Lake.

Now, go jump in the lake kid!

Nicole & Johanna

A is for an **Airboat**,
skimming across the bay.
It's like a magic carpet ride
that takes us out to play.

is for **Brine Shrimp** which is really hard to rhyme but if you look up close you can see their little beady eyes.

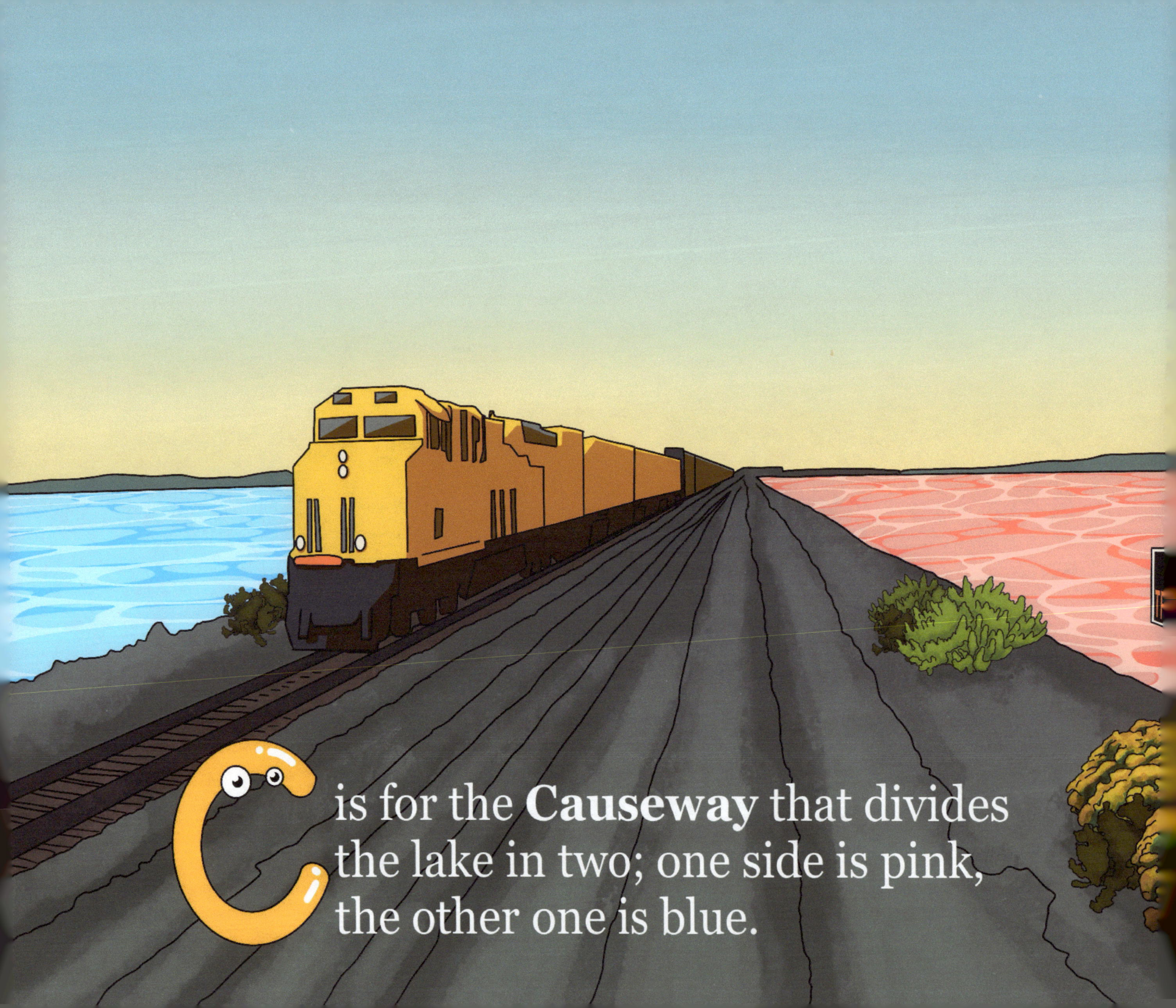

C is for the **Causeway** that divides the lake in two; one side is pink, the other one is blue.

D is for **Ducks** with webbed feet that dabble in the water while searching for a tasty treat.

E is for **Everything** that makes this lake so dear. Airboating, sailing, and duck hunters with long, funny beards.

F is for **Floating** in water you cannot drink, because it is so salty it will never let you sink.

G is for **Grebes** that nest throughout the marsh and fly south for winter when the weather gets too harsh.

H is for the **Horses** that cowboys like to ride; you can find them at the ranch on the island's eastern side.

I is for the **Islands** that dot our inland sea, providing rest and shelter for the birds, the beasts, and me.

J is for **John C. Fremont**, who stood high upon a peak, surveying the lake around him and all that he could see.

K is for **Kids** that play along the shore: running, splashing, and kayaking. Who could ask for more?

L is for **Legends** and tall tales galore about Great Salt Lake: water-babies, whales, monsters, and more.

M is for **Microbialites**, rocks that are alive. Keeping scientists busy unraveling the mystery of life.

N is for **Nighttime** when owls and stars appear; high up in the night sky, you can see the Milky Way so clear.

is **Orb Weavers,** who spin their webs with care, along the rocky shoreline and in the salty air.

P is for **Pelicans** soaring on the breeze, flying to Gunnison Island where they nest on our inland sea.

Q is for **Quiet,** a peaceful place to ponder. As I watch the sunset, I sit in hopeful wonder.

R is for the **Round-up** of Bison wild and free, a symbol of the west and what it used to be.

S is for the **Sailboats** that harness the briny breeze, sailing on the water of this great inland sea.

T is for the **Trains** that helped America unite when all the work was finished at the Golden Spike.

U is for **Underwater**, the place we want to be but because of lack of water, the land is all you see.

CAUTION

LOW WATER
LAUNCH AT
OWN RISK

V is for our **Voices** that must speak out as one to protect our lake, our Great Salt Lake, because it's the only one.

W is for the **Wetlands** found in Farmington Bay where ducks, egrets, and eagles, fly, and live, and play

X marks the spot, two **X's** and a dot. Fun, adventure, and play; floating or sailing away. Utah's best-kept secret is our Great Salt Lake.

Y is for the **Yearly migration** that thousands of birds must make, flying he to Utah to nest at the Great Salt Lake.

Z is for the **Zeal** that outdoorsmen must take, in protecting natural wonders, especially our Great Salt Lake.

Fun Facts

Airboats play an important role in search and rescue efforts if someone gets lost or hurt in the wetlands. Airboats have a flat underside, or hull, and have big motors with propellers - like an airplane - that allow the boat to push through tall reeds and phragmites where other boats can't go.

Brine shrimp, some no bigger than a pinhead, can control how much salt they allow into their bodies better than any other organisms in the world because of their skin lining and respiratory system that helps to remove extra salt from their bodies as they breathe in fresh oxygen.

Duck hunters provide a huge economic boost to Utah. According to a study by Dr. John Duffield, "Waterfowl hunting on and around the Great Salt Lake is responsible for an estimated $97 million in economic activity per year, supports 1,600 jobs, and provides nearly $37 million of income to Utah workers."

In 1893, the Mormon church built America's first amusement park, Saltair along the shoreline of the Great Salt Lake bringing over 10,000 visitors on its opening day.

Grebes eat their feathers to help protect their stomach from sharp food items they may find in the marsh. They also build a nest that floats and they carry their babies on their backs the first week of life.

The Fielding Garr Ranch on Antelope Island was established in 1848, one year after the Mormon Pioneers arrived in Salt Lake City. This makes it one of the oldest working ranches in the Western United States.

n C. Fremont and Kit Carson surveyed and conducted the first scientific study of the at Salt Lake in their homemade, "India Rubber" boat. The island was named after John Fremont to honor his achievement of mapmaking.

In the early 1800s, reports of mysterious creatures were part of everyday life: water-babies (mermaids) sunning themselves on the rocks, whales, and giants riding elephant-like creatures on the islands... all while the Great Salt Lake Monster ruled the waters of the lake. All fun but all fake!

obialites start as squishy mats of photosynthetic microbes and cyanobacteria that trap nent which binds to their mat, creating rocks that are alive. The presence of obialites at the Great Salt Lake is amazing as they represent some of the earliest forms e... some dating back to 3.4 billion years ago!

In 2017, Antelope Island earned the designationof an International Dark Sky Park for actively conserving its remaining darkness. The island is a great location for stargazing and night sky photography and has become a popular destination for area stargazers that want to get away from the glow of the city.

ern Spotted Orb Weavers like to hang around and catch brine flies, mosquitos, and s. They build circular, spiral wheel-shaped webs and they can grow as big as one to two es in diameter, not including their legs.

Gunnison Island, located in the northern arm of the lake, is home to one of North America's largest breeding colonies of American White Pelicans. Pelicans don't store their catch in their pouch as popular belief states but instead, they swallow and regurgitate it later. Pelicans also have a sharp hook (similar to a tooth) on the end of their beak.

The bison is the largest mammal in North America with bulls weighing up to 2,000 pounds and six feet tall. Bison are moody. If their tail is hanging down switching naturally, the bison is usually calm. If it's standing straight up... look out! Regardless of what a bison's tail is doing, remember they are unpredictable, huge, fast animals, and can charge at any moment. Give them their space.

The Union Pacific Railroad built westward and the Central Pacific Railroad built eastward until their rails joined together on May 10, 1869, at Promontory Point creating the first transcontinental railroad.

Only 1% of Utah consists of wetlands and approximately 80% of those wetlands are within the Great Salt Lake ecosystem. Wetlands have lots of different names; marsh, swamp, and mudflats. Their function is similar to a human kidney and works by absorbing waste such as nitrogen and phosphorus thus cleaning the water that flows through them.

The Great Salt Lake is an important stop on the Pacific Flyway; a highway in the sky for millions of birds on their annual migration from Alaska to Patagonia. The surrounding wetlands provide safety and snacks of brine flies that provide nutrients for the birds.

All species deserve a healthy habitat which are an underlying factor to survival in the wild. Many organizations work together to conserve and provide habitat for waterfowl populations; not only for hunting but also birdwatching and enjoyment for future generations.

Howdy There Kiddo!

People call me "the Lake" but my real name is Great Salt, the Great Salt Lake to be exact. I call the best state in the nation home; after all, Utah is one of the only places where you can go sailing and skiing on the same day!

I am a remnant of ancient Lake Bonneville and during my glory days of the Great Ice Age, I was HUGE! Literally HUGE! I measured in at 19,800 square miles, making me the largest Late Pleistocene paleolake in the Great Basin. About 30,000 years ago, I began to retreat and that's when people started calling me, the Great Salt Lake. I left my mark though. If you look toward the Wasatch Front you can see where my waves etched terraces into the mountainside.

I'm still a pretty big deal! I am the largest lake west of the Mississippi. I am fed water from three tributaries upstream; the Bear, Jordan, and Weber rivers but there aren't any rivers flowing out, away from my shoreline. I'm a terminal lake and what water I receive from the rivers eventually evaporates leaving behind trace minerals and lots of salt.

I'm much too salty for fish to survive and yet I am home to millions of tiny salt-loving single-cell organisms and brine shrimp. The water in my north arm - did I mention I had arms? Well, kind of anyway. I have four regions that people often describe as arms. So, the north arm. The water here is eight times saltier than the ocean, and it appears PINK – like pink lemonade – because of the pigments in these salt-lovers cells.

Within my arms, I hold many islands, animals, sand, and salt. Gunnison Island has been a sanctuary and home to the American White Pelican for decades. Most of the wetlands in Utah are along my eastern shoreline where millions of birds stopover for lunch and to take a nap during their annual migration from Alaska to Patagonia. Antelope Island, an International Dark Sky Park, is another remarkable location where they actively conserve its darkness.

And did I mention that I make my own sand? My sand is called oolitic sand and it is found in only a couple of other places on the planet. It has a nucleus – often brine shrimp poop – which is covered by a layer of calcium carbonate and then rolled along the lake bottom by my waves gradually accumulating layers.

These are just some of the awesome things that I hold in my arms. I like to think that I draw people to me like a magnet for a variety of recreational experiences and adventures. Hopefully to enjoy what John Muir called "one of the great views on the American Continent" and perhaps Utah's best-kept secret.

See you at the lake, kid!

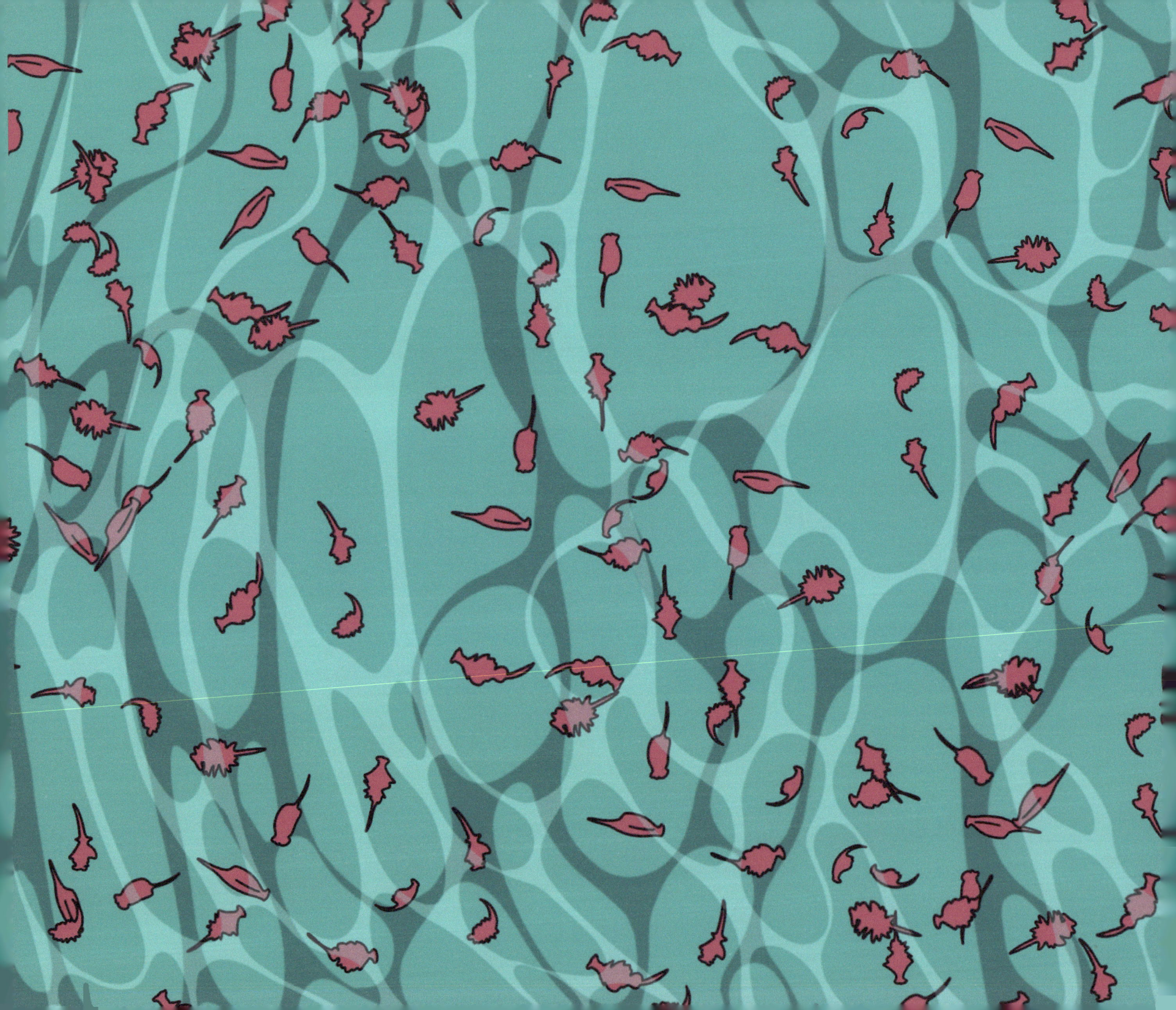

www.ingramcontent.com/pod-product-compliance
Lightning Source LLC
Chambersburg PA
CBHW041551030426

42335CB00004B/184